The Human Body Bingo Book

COMPLETE BINGO GAME IN A BOOK

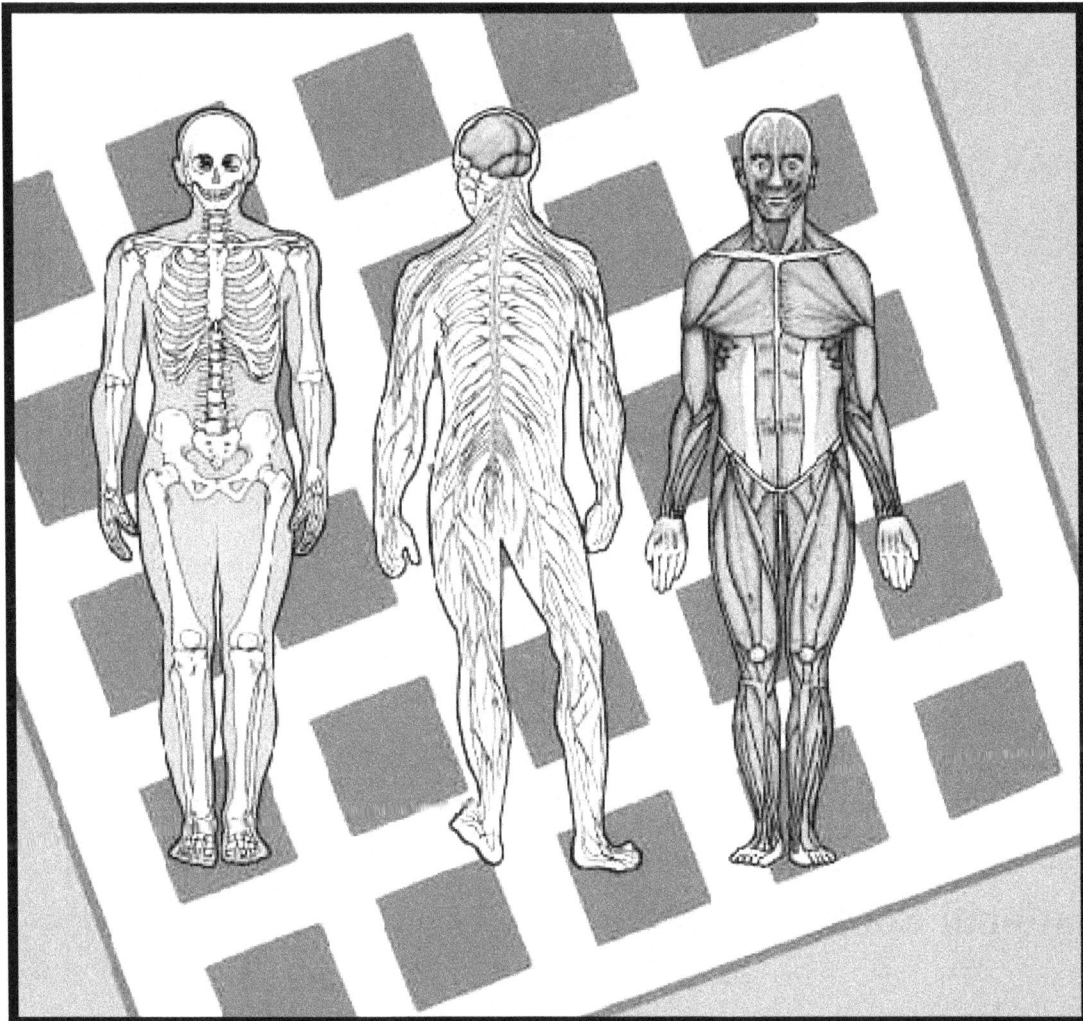

I0070802

Written By Rebecca Stark

ISBN 978-0-87386-437-4

Educational Books 'n' Bingo

Printed in the U.S.A.

THE HUMAN BODY BINGO
Directions

INCLUDED:

List of Terms

Templates for Additional Terms and Clues

2 Clues per Term

30 Unique Bingo Cards

Markers

1. **Either cut apart the book or make copies of ALL the sheets. You might want to make an extra copy of the clue sheets to use for introduction and review. Keep the sheets in an envelope for easy reuse.**

2. Cut apart the call cards with terms and clues.

3. Pass out one bingo card per student. There are enough for a class of 30.

4. Pass out markers. You may cut apart the markers included in this book or use any other small items of your choice.

5. Decide whether or not you will require the entire card to be filled. Requiring the entire card to be filled provides a better review. However, if you have a short time to fill, you may prefer to have them do the just the border or some other format. Tell the class before you begin what is required.

6. There are 50 topics. Read the list before you begin. If there are any topics that have not been covered in class, you may want to read to the students the topic and clues before you begin.

7. There is a blank space in the middle of each card. You can instruct the students to use it as a free space or you can write in answers to cover topics not included. Of course, in this case you would create your own clues. (Templates provided.)

8. Shuffle the cards and place them in a pile. Two or three clues are provided for each topic. If you plan to play the game with the same group more than once, you might want to choose a different clue for each game. If not, you may choose to use more than one clue.

9. Be sure to keep the cards you have used for the present game in a separate pile. When a student calls, "Bingo," he or she will have to verify that the correct answers are on his or her card AND that the markers were placed in response to the proper questions. Pull out the cards that are on the student's card keeping them in the order they were used in the game. Read each clue as it was given and ask the student to identify the correct answer from his or her card.

10. If the student has the correct answers on the card AND has shown that they were marked in response to the *correct questions,* then that student is the winner and the game is over. If the student does not have the correct answers on the card OR he or she marked the answers in response to *the wrong questions,* then the game continues until there is a proper winner.

11. If you want to play again, reshuffle the cards and begin again.

Have fun!

TERMS

artery (arteries)

atrium (atria)

blood

bone(s)

brain

bronchi

capillaries

cell(s)

cerebrum

circulatory

DNA

digestive

ear(s)

endocrine

esophagus

eye(s)

gene(s)

heart

hormone(s)

immune

intestine

joints

kidneys

larynx

lungs

mouth

muscles (muscular)

nerve

nervous

nose

organ(s)

oxygen

pulmonary

reflex

respiratory

senses

skeletal

skull

spinal cord

stomach

system(s)

teeth

tissue(s)

torso

trachea

urinary

veins

ventricle

vertebra(ae)

x-ray(s)

© **Barbara M. Peller**

Additional Terms

Choose as many terms as you would like and write them in the squares. Repeat each as desired. Cut out the squares and randomly distribute them to the class. Instruct the students to place the square on the center space of their card.

The Human Body Bingo

Clues for Additional Terms

Write two clues for each of your additonal terms.

1. 2.	1. 2.
1. 2.	1. 2.
1. 2.	1. 2.

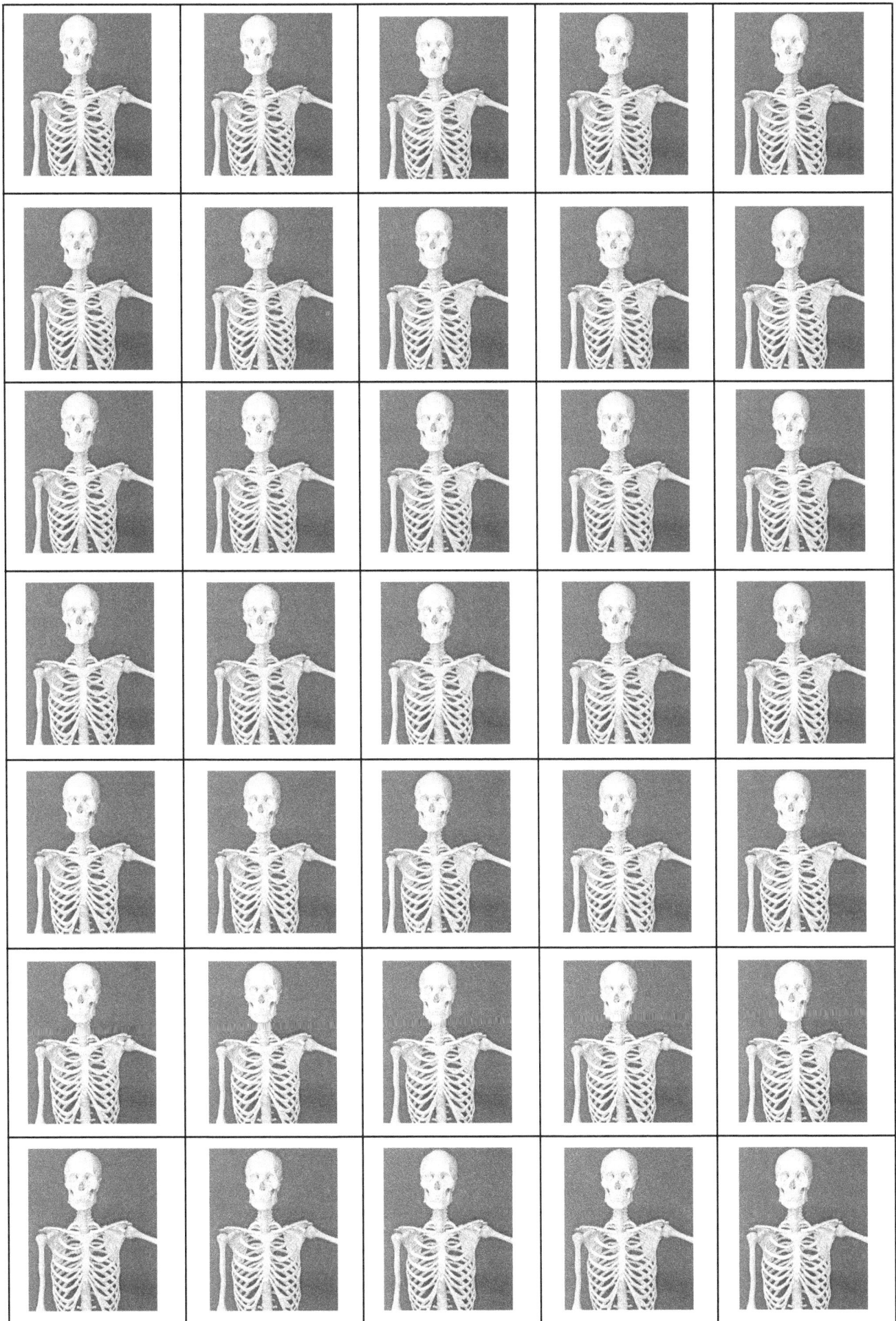

artery (arteries)	**atrium (atria)**
1. These large blood vessels look red and carry blood away from the heart. 2. The largest ___ in the human body is the aorta.	1. The ___ are the upper chambers of the heart. They receive blood from the veins and push it into the ventricles. 2. The right ___ pushes blood into the right ventricle and the left ___ pushes blood into the left ventricle.
blood	**bone(s)**
1. The fluid that is circulated through the body by the heart is called ___. 2. ___ contains plasma, ___ cells, and platelets.	1. The adult human body has 206 ___. 2. The smallest ___ is the stirrup, which carries sound from the eardrum to the inner ear.
brain	**bronchi**
1. This organ controls everything we do. The part that controls thinking is the cerebrum. 2. The main parts of the ___ are the cerebrum, the cerebellum, and the ___ stem.	1. These tubes are part of the respiratory system. They branch off from the trachea and connect it to the lungs. 2. The ___ convey air to and from the lungs.
capillaries	**cell(s)**
1. The smallest blood vessels are called ___. 2. These tiny blood vessels connect the smallest arteries, called arterioles, with the smallest veins, called venules.	1. The ___ is the basic structural and functional unit of all living things. The adult human has about 100 trillion of them! 2. A tissue is a group of ___ with a similar function and structure.
cerebrum	**circulatory**
1. The ___ fills most of the cranial cavity. It controls thought as well as motor and sensory functions. 2. The ___ is divided into two hemispheres; these hemispheres are joined at the bottom by the corpus callosum.	1. This system comprises the heart, arteries, capillaries and veins. 2. The ___ system controls the flow of blood around the body.

The Human Body Bingo

DNA 1. Its real name is deoxyribonucleic acid and it contains our genetic codes. 2. Lengths of connected ___ molecules are called genes.	**digestive** 1. Organs in this system include the mouth, the esophagus, the stomach, the small intestine, and the large intestine, also called the colon. 2. Organs in this system help the body break down and absorb food.
ear(s) 1. This organ detects, processes and sends sound signals to your brain. It has three sections. 2. Your ___ not only allow you to hear and process sounds, but also help you keep your balance.	**endocrine** 1. The ___ system includes the pituitary, the thyroid and other glands. The glands secrete chemical substances known as hormones. 2. The ___ system helps regulate metabolism, growth, development and tissue function. It also affects a person's mood.
esophagus 1. The tube that connects the pharynx, or throat, with the stomach is called the ___. 2. When you swallow, the muscular walls of your ___ contract, pushing the food down into the stomach.	**eye(s)** 1. These are the organs of sight. 2. The sclera, cornea, retina, lens, iris and pupil are all parts of the ___.
gene(s) 1. A ___ is the basic unit of heredity. 2. This unit of heredity is a piece of DNA that is passed on from parent to offspring.	**heart** 1. This muscle sends blood around the body. 2. The word cardiac means "of or related to the ___."
hormone(s) 1. ___ are chemical substances produced in the body. They control and regulate the activity of certain cells or organs. 2. The pancreas secretes a ___ called Insulin.	**immune** 1. The ___ system helps keep out harmful bacteria and viruses. It attacks those that get inside your body. 2. Two signs that your ___ system is working are getting better after being sick and cuts healing without getting infected.

The Human Body Bingo

intestine 1. The ___ is part of the digestive system. There are the two major sections: the small ___ and the large ___. 2. The small ___ is about 20 feet long. The large ___, also called the colon, is wider but only about 5 feet long.	**joints** 1. Hinge ___ allow the body to bend and straighten. The elbow, knee and ankle are examples. 2. The shoulder and hip are ball-and-socket ___. Ball-and-socket ___ are the most flexible.
kidneys 1. These bean-shaped organs are the most important organs of the urinary system. Most people have two. 2. The ___ filter wastes from the blood and excrete them and water in urine.	**larynx** 1. This organ is sometimes called the voice box. 2. The ___ is part of the respiratory tract. It contains the vocal cords, which produce sound.
lungs 1. They are our basic respiratory organs. We each have a pair of them in our chest. 2. These breathing organs remove carbon dioxide from the blood and bring oxygen to the blood.	**mouth** 1. It is the upper opening of the digestive tract. 2. Digestion begins in the ___, where saliva starts to break down the chemicals in the food.
muscles (muscular) 1. There are three types of ___ in the ___ system: smooth, cardiac, and skeletal. 2. Skeletal ___ are called voluntary because we can control what they do. Smooth ___ are called involuntary.	**nerve** 1. A ___ is a bundle of fibers that transmits information from one body part to another. 2. ___ cells are called neurons. Neurons have specialized extensions called dendrites and axons.
nervous 1. The brain and the spinal cord make up the central ___ system. 2. The peripheral ___ system includes the nerves and ganglia.	**nose** 1. This organ of smell is part of the respiratory system. 2. The two holes in the ___ are called nostrils.

The Human Body Bingo   © Barbara M. Peller

organ(s) 1. An ___ is a group of tissues that performs one or more specific functions. 2. The eyes, ears, kidneys, heart and lungs are all ___ of the human body.	**oxygen** 1. The main function of the respiratory system is to allow ___ in the air to be taken into the body. 2. ___ is taken into the body and carbon dioxide is breathed out of the body.
pulmonary 1. The ___ arteries carry blood from the heart to the lungs. 2. Complete this analogy: cardiac : heart :: ___ : lungs.	**reflex** 1. A ___ is an involuntary reaction. 2. An involuntary response, such as a sneeze or a hiccup, is called a ___.
respiratory 1. The main organs of the ___ system are the lungs, the trachea, the bronchi, and the diaphragm. 2. The organs of the ___ system help us breathe.	**senses** 1. The specialized functions by which we receive and respond to stimuli are called ___. 2. Our ___ include sight, hearing, smell, taste and touch.
skeletal 1. The ___ system supports and protects our bodies. It is sometimes called the musculo-___ system. 2. This system comprises the bones and the tissues that connect them, including tendons, ligaments and cartilage.	**skull** 1. Your ___ is part of the skeletal system. It protects your brain and eyes. 2. The ___ is also called a cranium.
spinal cord 1. The ___ is the thin, tubular bundle of nervous tissue that extends from the brain and continues along the back. 2. The ___ is protected by the spinal column, which is made of bony segments called vertebrae. The Human Body Bingo	**stomach** 1. This sac-like organ is the main organ of digestion. 2. This organ of the digestive system is between the esophagus and the small intestine. © Barbara M. Peller

system(s) 1. A ___ is made up of organs and related tissues concerned with the same function. 2. Some ___ of the human body include the nervous ___, the respiratory ___, the reproductive ___, and the circulatory ___.	**teeth** 1. We use them to bite and chew food. They are part of the skeletal system but are not bones. 2. ___ are made mostly of dentin. Enamel, the outer layer, is the hardest substance in the body.
tissue(s) 1. A group of cells that perform a specific function is called a ___. 2. Cells are organized into ___; ___ are organized into organs; and organs are organized into systems.	**torso** 1. The ___ refers to the body except for the head, neck and limbs. 2. Your arms and legs are parts of your body, but they are not parts of your ___.
trachea 1. The ___ is sometimes called the windpipe. It branches off into two bronchi. 2. This tube is the main airway through which air passes to and from the lungs.	**urinary** 1. The ___ system is an excretory system. It rids your body of toxic waste. 2. Major organs of the ___ system are the kidneys, the bladder, the ureters, and the urethra.
veins 1. These blood vessels carry blood toward the heart. 2. They are similar to arteries, but they are not as strong.	**ventricle** 1. The lower two chambers of the heart are called the left ___ and the right ___. 2. The right ___ pumps blood to the pulmonary artery. The left ___ pumps blood to the aorta.
vertebra(ae) 1. A ___ is one of the bony segments of the spinal column. 2. The spinal column is made up of 33 ___.	**x-ray(s)** 1. ___ can be used to take pictures of the inside of the body. 2. An image produced on photographic film by ___ passing through parts of the body is sometimes used as a diagnostic tool.
The Human Body Bingo	

The Human Body Bingo

heart	blood	joints	veins	teeth
cerebrum	hormone(s)	trachea	nerve	mouth
spinal cord	pulmonary		intestine	nose
urinary	atrium (atria)	gene(s)	x-ray(s)	kidneys
larynx	vertebra(ae)	DNA	artery (arteries)	immune

The Human Body Bingo

veins	system(s)	lungs	oxygen	larynx
kidneys	esophagus	cell(s)	atrium (atria)	skull
respiratory	vertebra(ae)		digestive	gene(s)
nerve	senses	pulmonary	ventricle	mouth
immune	trachea	DNA	cerebrum	artery (arteries)

The Human Body Bingo

veins	gene(s)	nerve	x-ray(s)	spinal cord
vertebra(ae)	blood	bronchi	hormone(s)	nervous
atrium (atria)	trachea		skeletal	bone(s)
pulmonary	respiratory	larynx	esophagus	lungs
artery (arteries)	DNA	cerebrum	ventricle	joints

The Human Body Bingo

pulmonary	skeletal	joints	DNA	larynx
muscles (muscular)	esophagus	hormone(s)	oxygen	spinal cord
intestine	cell(s)		teeth	x-ray(s)
gene(s)	eye(s)	trachea	cerebrum	bronchi
artery (arteries)	immune	organ(s)	endocrine	nose

The Human Body Bingo: Card No. 4

The Human Body Bingo

immune	teeth	atrium (atria)	cell(s)	DNA
muscles (muscular)	gene(s)	bronchi	pulmonary	ear(s)
system(s)	nose		blood	joints
mouth	skeletal	heart	ventricle	endocrine
nerve	cerebrum	reflex	digestive	intestine

The Human Body Bingo

bone(s)	skeletal	lungs	system(s)	nose
x-ray(s)	atrium (atria)	endocrine	hormone(s)	spinal cord
oxygen	bronchi		cell(s)	digestive
cerebrum	larynx	ventricle	organ(s)	intestine
kidneys	gene(s)	heart	reflex	joints

The Human Body Bingo

heart	skeletal	stomach	ear(s)	nerve
kidneys	joints	vertebra(ae)	blood	spinal cord
lungs	x-ray(s)		digestive	brain
pulmonary	esophagus	muscles (muscular)	veins	respiratory
DNA	cerebrum	ventricle	organ(s)	bone(s)

The Human Body Bingo: Card No. 7

The Human Body Bingo

intestine	skeletal	capillaries	x-ray(s)	brain
muscles (muscular)	system(s)	oxygen	joints	teeth
spinal cord	skull		nose	cell(s)
artery (arteries)	pulmonary	veins	endocrine	esophagus
trachea	cerebrum	organ(s)	atrium (atria)	kidneys

The Human Body Bingo

digestive	nerve	vertebra(ae)	spinal cord	nose
endocrine	system(s)	intestine	atrium (atria)	joints
nervous	heart		blood	capillaries
brain	immune	larynx	ear(s)	stomach
esophagus	ventricle	bronchi	veins	teeth

The Human Body Bingo

urinary	veins	cell(s)	oxygen	reflex
nose	brain	hormone(s)	blood	joints
skeletal	skull		x-ray(s)	respiratory
larynx	mouth	endocrine	ventricle	nervous
circulatory	immune	lungs	kidneys	intestine

The Human Body Bingo

bone(s)	skull	atrium (atria)	endocrine	kidneys
capillaries	nervous	ear(s)	digestive	hormone(s)
muscles (muscular)	system(s)		lungs	vertebra(ae)
circulatory	spinal cord	ventricle	cerebrum	veins
bronchi	DNA	heart	organ(s)	nerve

The Human Body Bingo

nerve	esophagus	nervous	x-ray(s)	digestive
vertebra(ae)	trachea	system(s)	organ(s)	muscles (muscular)
heart	stomach		nose	oxygen
DNA	teeth	joints	veins	blood
skull	capillaries	skeletal	bronchi	brain

The Human Body Bingo

circulatory	teeth	bone(s)	nervous	nose
system(s)	capillaries	skeletal	digestive	respiratory
x-ray(s)	cell(s)		vertebra(ae)	stomach
intestine	ventricle	brain	skull	veins
cerebrum	mouth	organ(s)	heart	ear(s)

The Human Body Bingo

DNA	system(s)	atrium (atria)	digestive	circulatory
brain	heart	nervous	blood	respiratory
endocrine	x-ray(s)		lungs	cell(s)
mouth	ventricle	skeletal	bronchi	bone(s)
cerebrum	oxygen	skull	kidneys	intestine

The Human Body Bingo: Card No. 14

The Human Body Bingo

ear(s)	digestive	atrium (atria)	nerve	joints
bone(s)	reflex	hormone(s)	system(s)	endocrine
nose	heart		spinal cord	x-ray(s)
cerebrum	nervous	capillaries	ventricle	circulatory
kidneys	esophagus	organ(s)	lungs	vertebra(ae)

The Human Body Bingo

cell(s)	torso	capillaries	reflex	senses
oxygen	skull	stomach	muscles (muscular)	urinary
circulatory	teeth		nose	vertebra(ae)
pulmonary	esophagus	cerebrum	ear(s)	veins
endocrine	nervous	organ(s)	brain	respiratory

The Human Body Bingo

circulatory	tissue(s)	eye(s)	nervous	cerebrum
ear(s)	endocrine	ventricle	x-ray(s)	stomach
digestive	urinary		torso	capillaries
immune	kidneys	intestine	atrium (atria)	respiratory
larynx	bronchi	nerve	veins	teeth

The Human Body Bingo

joints	skeletal	brain	endocrine	oxygen
immune	circulatory	atrium (atria)	nose	bronchi
digestive	respiratory		eye(s)	reflex
skull	hormone(s)	ventricle	urinary	lungs
torso	nervous	larynx	tissue(s)	bone(s)

The Human Body Bingo: Card No. 18

The Human Body Bingo

nose	bone(s)	nervous	capillaries	skull
ear(s)	DNA	reflex	nerve	urinary
tissue(s)	x-ray(s)		blood	atrium (atria)
lungs	torso	larynx	esophagus	eye(s)
spinal cord	senses	kidneys	intestine	organ(s)

The Human Body Bingo

skull	tissue(s)	urinary	nervous	blood
cell(s)	vertebra(ae)	muscles (muscular)	larynx	oxygen
teeth	stomach		pulmonary	eye(s)
immune	intestine	artery (arteries)	esophagus	torso
gene(s)	trachea	senses	veins	hormone(s)

The Human Body Bingo

bone(s)	immune	muscles (muscular)	nervous	mouth
teeth	eye(s)	brain	capillaries	heart
respiratory	kidneys		tissue(s)	atrium (atria)
larynx	nerve	torso	ear(s)	intestine
pulmonary	senses	organ(s)	circulatory	esophagus

The Human Body Bingo

spinal cord	lungs	eye(s)	system(s)	circulatory
oxygen	urinary	joints	capillaries	blood
brain	x-ray(s)		heart	stomach
torso	immune	esophagus	hormone(s)	DNA
senses	bronchi	tissue(s)	respiratory	muscles (muscular)

The Human Body Bingo

cell(s)	tissue(s)	nerve	system(s)	organ(s)
bone(s)	skull	kidneys	ear(s)	hormone(s)
lungs	circulatory		artery (arteries)	heart
respiratory	trachea	torso	bronchi	esophagus
mouth	intestine	senses	larynx	eye(s)

The Human Body Bingo

cell(s)	skull	DNA	tissue(s)	capillaries
nose	organ(s)	muscles (muscular)	oxygen	heart
stomach	reflex		circulatory	respiratory
mouth	artery (arteries)	torso	bronchi	teeth
gene(s)	pulmonary	senses	urinary	trachea

The Human Body Bingo: Card No. 24

© Barbara M. Peller

The Human Body Bingo

pulmonary	muscles (muscular)	tissue(s)	atrium (atria)	eye(s)
hormone(s)	mouth	ear(s)	cell(s)	blood
teeth	capillaries		artery (arteries)	torso
reflex	immune	trachea	senses	urinary
organ(s)	DNA	brain	endocrine	gene(s)

The Human Body Bingo

eye(s)	tissue(s)	artery (arteries)	oxygen	reflex
lungs	x-ray(s)	capillaries	skull	cell(s)
mouth	larynx		urinary	pulmonary
circulatory	system(s)	immune	senses	torso
stomach	endocrine	atrium (atria)	trachea	gene(s)

The Human Body Bingo

artery (arteries)	brain	tissue(s)	skull	vertebra(ae)
mouth	lungs	ear(s)	torso	blood
ventricle	trachea		senses	pulmonary
reflex	bone(s)	gene(s)	muscles (muscular)	hormone(s)
circulatory	urinary	eye(s)	spinal cord	stomach

The Human Body Bingo

nose	skeletal	veins	tissue(s)	brain
vertebra(ae)	eye(s)	artery (arteries)	larynx	urinary
trachea	respiratory		reflex	oxygen
stomach	spinal cord	kidneys	senses	torso
system(s)	digestive	circulatory	gene(s)	mouth

The Human Body Bingo: Card No. 28

The Human Body Bingo

eye(s)	skeletal	reflex	ear(s)	digestive
mouth	larynx	muscles (muscular)	stomach	spinal cord
teeth	tissue(s)		blood	artery (arteries)
vertebra(ae)	immune	joints	senses	torso
cell(s)	capillaries	gene(s)	bone(s)	trachea

The Human Body Bingo: Card No. 29

The Human Body Bingo

DNA	tissue(s)	oxygen	digestive	torso
hormone(s)	reflex	lungs	urinary	blood
gene(s)	trachea		stomach	muscles (muscular)
mouth	bone(s)	eye(s)	senses	artery (arteries)
immune	nerve	bronchi	skeletal	joints

The Human Body Bingo: Card No. 30

www.ingramcontent.com/pod-product-compliance
Lightning Source LLC
Chambersburg PA
CBHW051420200326
41520CB00023B/7306